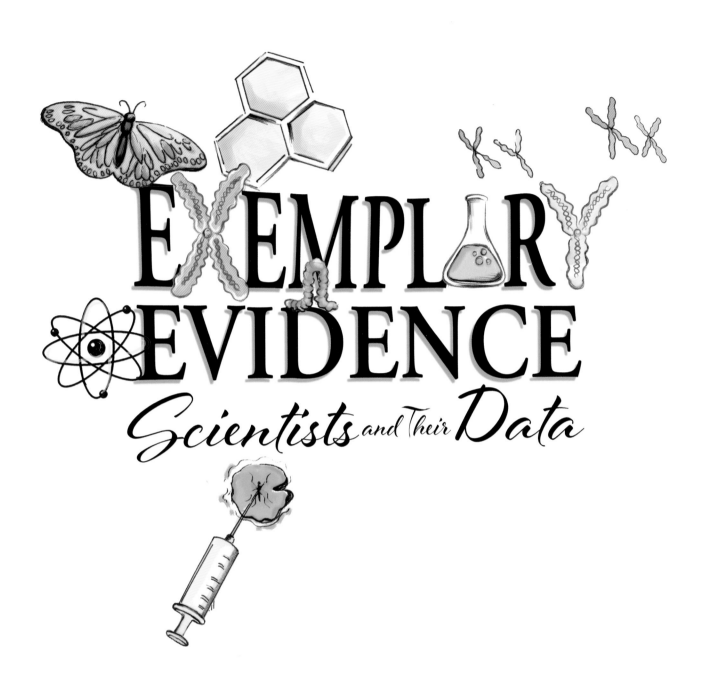

EXEMPLARY EVIDENCE

Scientists and Their Data

National Science Teachers Association

Claire Reinburg, Director
Rachel Ledbetter, Managing Editor
Deborah Siegel, Associate Editor
Andrea Silen, Associate Editor
Donna Yudkin, Book Acquisitions Manager

ART AND DESIGN
Will Thomas Jr., Director
Linda Olliver, Cover, Interior Design
Illustrations by Linda Olliver

PRINTING AND PRODUCTION
Catherine Lorrain, Director

NATIONAL SCIENCE TEACHERS ASSOCIATION
David L. Evans, Executive Director

1840 Wilson Blvd., Arlington, VA 22201
www.nsta.org/store
For customer service inquiries, please call 800-277-5300.

Lexile® measure: 990L

NSTA is committed to publishing material that promotes the best in inquiry-based science education. However, conditions of actual use may vary, and the safety procedures and practices described in this book are intended to serve only as a guide. Additional precautionary measures may be required. NSTA and the authors do not warrant or represent that the procedures and practices in this book meet any safety code or standard of federal, state, or local regulations. NSTA and the authors disclaim any liability for personal injury or damage to property arising out of or relating to the use of this book, including any of the recommendations, instructions, or materials contained therein.

Library of Congress Cataloging-in-Publication Data

Names: Fries-Gaither, Jessica, 1977- author. | Olliver, Linda, illustrator.
Title: Exemplary evidence : scientists and their data / by Jessica
    Fries-Gaither ; illustrated by Linda Olliver.
Description: Arlington, VA : National Science Teachers Association, [2018] |
    Audience: Ages 8-10. | Audience: Grades 4 to 6.
Identifiers: LCCN 2018013924 (print) | LCCN 2018017550 (ebook) | ISBN
    9781681403625 (e-book) | ISBN 9781681403618 (print) | ISBN 9781681406558 (hard cover)
Subjects: LCSH: Science--Methodology--Juvenile literature. |
    Science--History--Juvenile literature. | Scientists--Juvenile literature.
Classification: LCC Q175.2 (ebook) | LCC Q175.2 .F74945 2018 (print) | DDC
    507.2/1--dc23
LC record available at https://lccn.loc.gov/2018013924

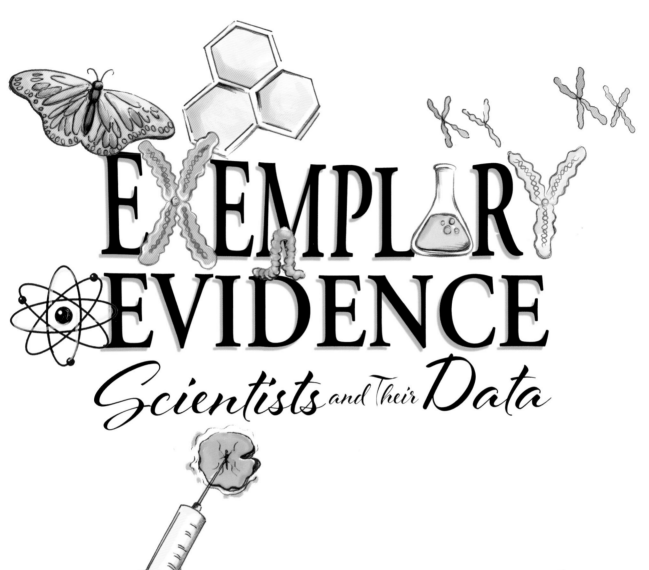

EXEMPLARY EVIDENCE

Scientists and Their Data

By **Jessica Fries-Gaither**

Illustrated by **Linda Olliver**

NSTA **Kids**

National Science Teachers Association

Arlington, VA

Scientists ponder, question, and wonder
about all kinds of subjects, from flowers to thunder.
But no matter what the query's about,
a scientist's job is to figure it out.

How do scientists find out what is true?
They need to have data—and lots of it, too!
Data can be tallies, measurements, numbers,
or notes and sketches of amazing wonders.

You can gather data through experimentation,
a process that can lead to much calculation.
So measurements must be precisely taken.
When it comes to data—you can't be mistaken.

How far did that travel? How fast did that go?
Answering these questions gives data, you know.
Timers and rulers, balances and more,
data is collected with tools galore.

Afterward, there's just one thing to do.
Analyze the data to find something new
by comparing, graphing, looking for trends.
What will it tell? Well, that depends.

How Much Ice Melted on Marble in 30 Minutes

Ice Cube Number	Weight (mL)
1	14
2	14.3
3	13.7
4	15.3
5	14.7

	1	2	3	4	5
GROUND COVER	GRASS AND OTHER PLANTS	ASPHALT	GRAVEL	TREES AND GRASS	ASPHALT
SUN/SHADE	SUN	SHADE	SUN	SHADE	SUN
AVERAGE TEMPERATURE	25.5 C	28.9 C	38.6 C	22.4 C	39.2 C

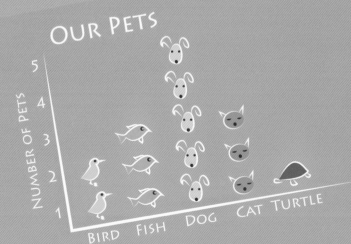

No matter the data, there is just one aim—
gather enough to make a strong claim.
Data is evidence for what's thought to be true.
It is the foundation of what scientists do.

The importance of data is shown rather well
by the work of scientists—too many to tell!
Throughout all of history, data's been key
in the making of every great discovery.

BIRDS THAT VISITED OUR FEEDER

SPARROW 20%

BLUE JAY 20%

CHICKADEE 25%

GRACKLE 5%

CARDINAL 30%

BUTTERFLY POPULATION IN MY BACK YARD

AREA OCCUPIED (SQ FT)

16
12
8
4
0

2012   2014   2016   2018

YEAR

OUR PETS

NUMBER OF PETS

5
4
3
2
1

BIRD   FISH   DOG   CAT   TURTLE

Alhazen lived in Egypt a thousand years ago,
yet still is important for what he did show.
A groundbreaking study not hard to understand;
investigations need not always be grand.

Into a darkened room, two lanterns shone light,
each of them hung from a different height.
On the far wall, two bright spots could be seen,
each of them formed by a lantern's strong beam.

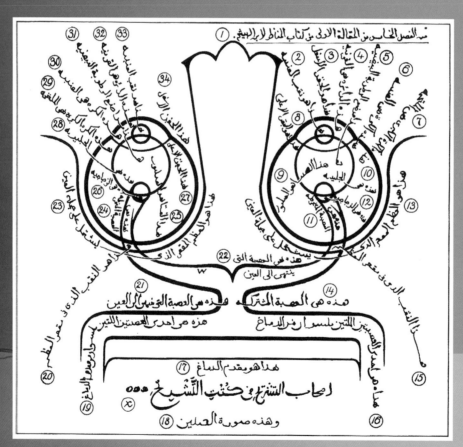

Alhazen's schematic of the visual system

8

If he covered a lantern, the spot became dark.
This simple finding ignited a spark.
Light did not come from our eyes, as believed!
Instead, it came from a source, he perceived.

Alhazen went on to more complex topics
and even wrote seven books about optics.
He is remembered, as his method was clear.
He was the first to use data to support
his idea.

The power to show that an idea is untrue—
it's amazing what **collecting data** can do.

Maria Merian's data took another form;
in fact, her life and work fell well outside the norm.
During her life, she painted insects of all kinds,
traveling far away to see what she could find.

Not much was known about insects or how they grew.
Where did they come from? The answer, no one knew.
Some popular ideas included mud, air, or rain.
Exactly how this happened, no one could explain.

Merian kept caterpillars; she watched them grow and change
even though to others, her studies seemed quite strange.
Through her colorful paintings, she could document
stages of a life cycle, and what those changes meant.

Merian's work was remarkably complete;
to show such detail was really quite a feat.
This precision helped other people see
the lives of insects, from butterflies to bees.

The power to develop a new point of view—
it's amazing what **recording data** can do.

One of Maria Merian's paintings

In the 1800s, new elements were found;
these discoveries seemed to be all around.
Chemist Dimitri Mendeleev had a gut feeling
that organizing their data would be quite revealing.

Upon cards, he wrote each one's information
and then tried his hand at organization.
After many hours, he fell asleep at his desk,
but when he awoke, he knew what to do next.

Reihen	Gruppe I. — R²O	Gruppe II. — RO	Gruppe III. — R²O³	Gruppe IV. RH⁴ RO²	Gruppe V. RH³ R²O⁵	Gruppe VI. RH² RO³	Gruppe VII. RH R²O⁷	Gruppe VIII. — RO⁴
1	H=1							
2	Li=7	Be=9.4	B=11	C=12	N=14	O=16	F=19	
3	Na=23	Mg=24	Al=27.3	Si=28	P=31	S=32	Cl=35.5	
4	K=39	Ca=40	—=44	Ti=48	V=51	Cr=52	Mn=55	Fe=56, Co=59, Ni=59, Cu=63.
5	(Cu=63)	Zn=65	—=68	—=72	As=75	Se=78	Br=80	
6	Rb=85	Sr=87	?Yt=88	Zr=90	Nb=94	Mo=96	—=100	04, Rh=104, 6, Ag=108.
7	(Ag=108)	Cd=112	In=113	Sn=118	Sb=122	Te=	J=	
8	Cs=133	Ba=137	?Di=138	?Ce=140	—	—	—	
9	(—)	—	—	—	—	—	—	
10	—	—	?Er=178	?La=180	Ta=182	W=184	—	197, 99.
11	(Au=199)	Hg=200	Tl=204	Pb=207	Bi=208	—	—	
12	—	—	—	Th=231	—	U=240	—	

He arranged the elements in order by weight
and knew he was onto something quite great.
Chemistry was changed by this development,
Mendeleev's Periodic Table of Elements.

The table wasn't just a success on its own;
it predicted elements before they were known.
It also pointed out some errors in weight.
Fortunately, the records could now be set straight.

The power to predict something entirely new —
it's amazing what **organizing data** can do.

In Havana, Cuba, the situation was dire;
deadly yellow fever was spreading like fire.
Outbreaks occurred time and again;
if only doctors could predict where and when.

Carlos Juan Finlay was up to the task.
As a family doctor, he just had to ask.
His patients' histories provided him clues;
the data was easy to collect and to use.

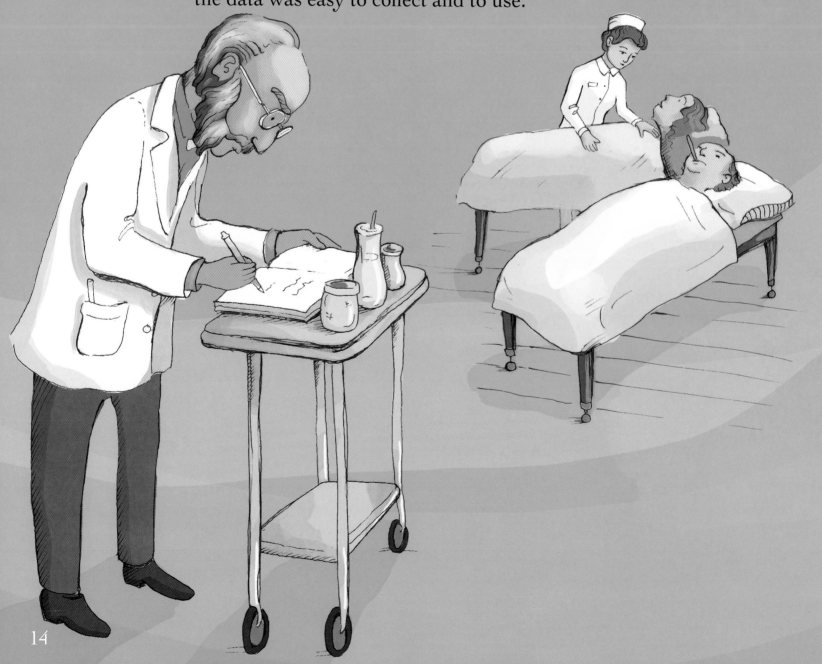

He looked for patterns, and with good reason;
most outbreaks occurred in the same season.
From his data he inferred the link:
Mosquitos spread the disease in a blink.

Sadly, his ideas weren't accepted for years:
The mosquito hypothesis fell on deaf ears.
Finlay persisted until others agreed
that mosquito prevention was a definite need.

The power to stop diseases and prevent them, too—
it's amazing what **comparing data** can do.

Nettie Stevens studied mealworms in great detail
to find out what made them male and female.
By using a microscope to observe DNA,
She collected data we use still today.

Peering through her microscope, Dr. Stevens saw
two different chromosomes: one short, the other tall.
Male mealworms consisted of one of each kind;
females had two tall ones—a surprising find.

It was the chromosomes! Dr. Stevens did declare.
The sex of the larva was determined by the pair.
Two of the same meant girl mealworms would be seen.
One of each? Boy mealworms, determined by their genes.

This work was the first to link chromosomes to traits.
Her clear observations would set the books straight.
All along, data hidden from the naked eye
showed females are XX, and males are XY.

The power to understand what's out of view—
it's amazing what **sharing data** can do.

Dr. Ruby Hirose, a chemist, had a brilliant mind.
Her data was of a quite different kind.
She conducted experiments on blood and cells,
collecting data to help keep us well.

She suffered from an allergy known as hay fever
and studied ways to improve a reliever.
Adding a new substance to make it work well
was a discovery of which others would tell.

She contributed to other studies, too;
a vaccine was the goal for the work that she'd do.
To prevent a disease called infantile paralysis,
her data would need some careful analysis.

Dr. Hirose's work did not go unseen,
such important studies being anything but routine.
At a meeting of chemists of great acclaim,
she was one of ten women mentioned by name.

The power to improve lives through and through —
it's amazing what **interpreting data** will do.

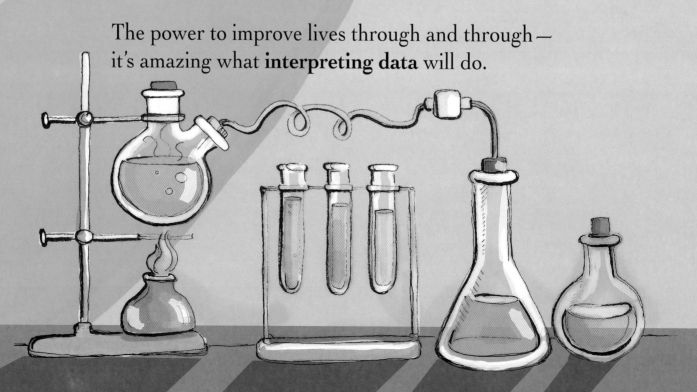

Dr. Ruby Hirose working in her lab

Down deep in the ocean, at the bottom of the sea,
Marie Tharp wondered what there might be.
Very few scientists cared about that,
for they assumed it was boring and flat.

No one could travel there, she had to concede.
So how could they get the data they'd need?
Sonar was used—a great innovation
to measure the depth of underwater locations.

She couldn't go to sea; she stayed on dry land,
plotting out thousands of soundings by hand.
Tedious work, but she and her partner pressed on,
consulting the data for what should be drawn.

Mountain ranges and valleys began to take shape;
it was an entire underwater landscape!
The map she created of the whole ocean floor
was used to prove others' theories and more.

The power to change our current world view—
it's amazing what **visualizing data** can do.

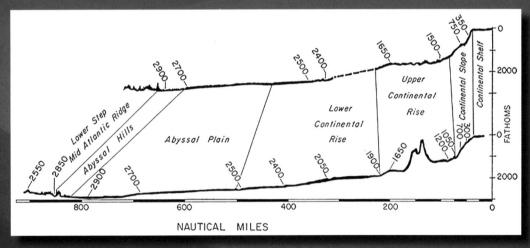

Map of the ocean floor created using Marie Tharp's sounding data

21

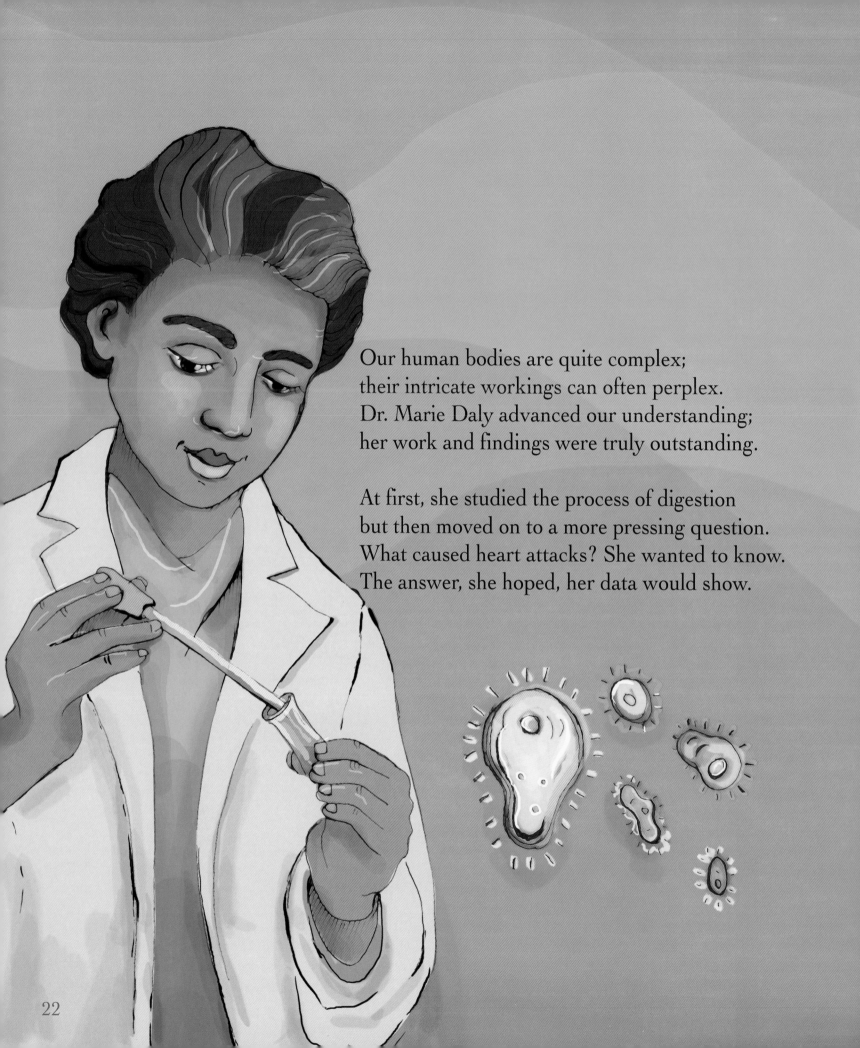

Our human bodies are quite complex;
their intricate workings can often perplex.
Dr. Marie Daly advanced our understanding;
her work and findings were truly outstanding.

At first, she studied the process of digestion
but then moved on to a more pressing question.
What caused heart attacks? She wanted to know.
The answer, she hoped, her data would show.

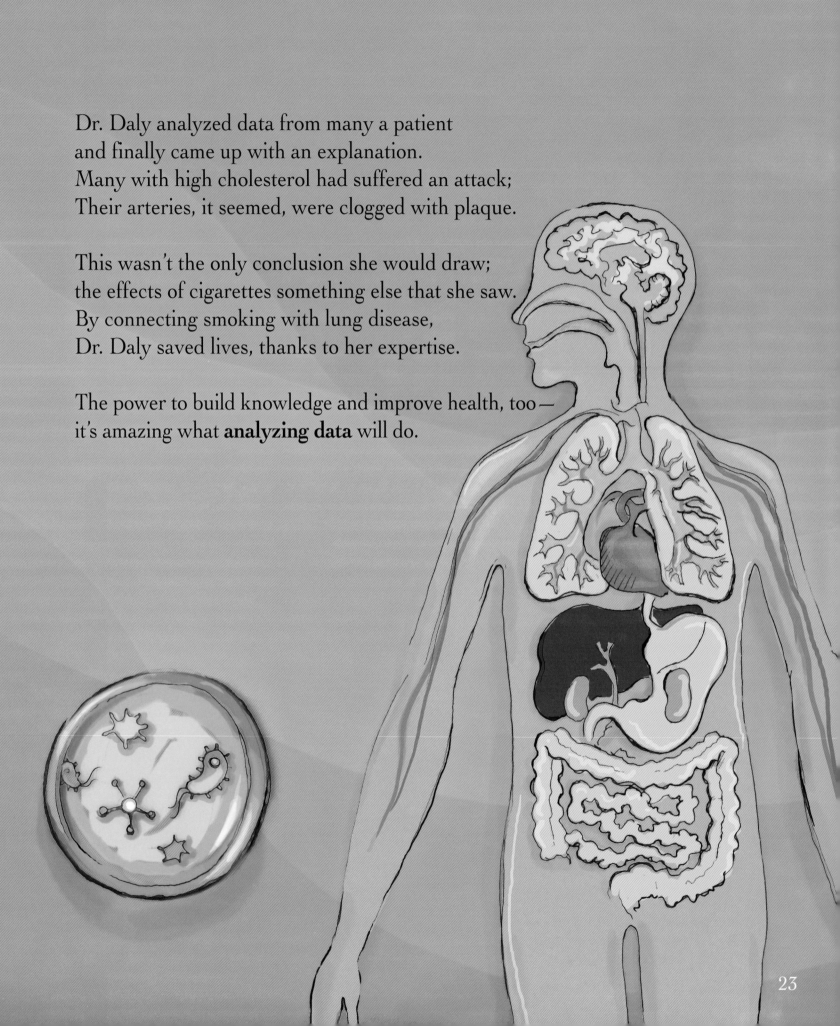

Dr. Daly analyzed data from many a patient
and finally came up with an explanation.
Many with high cholesterol had suffered an attack;
Their arteries, it seemed, were clogged with plaque.

This wasn't the only conclusion she would draw;
the effects of cigarettes something else that she saw.
By connecting smoking with lung disease,
Dr. Daly saved lives, thanks to her expertise.

The power to build knowledge and improve health, too—
it's amazing what **analyzing data** will do.

23

In places where coal, silver, and oil are found,
these precious resources are under the ground.
How do you know this when you can't see?
By using data to predict where they'll be.

A geologist in Montana can make such a claim.
Dr. Russell Stands-Over-Bull is his full name.
Born and raised a proud member of the Crow Nation,
his expertise lies in mapping resource locations.

Great
Falls Field

Bull Mountain

Fort
Union
Region

MONTANA

BITUMINOUS COAL
More than 14" Thick
Less than 14" Thick
Known Occurrence of Lignite in Tertiary Lake Beds

SUB-BITUMINOUS COAL
More Than 30" Thick
Less Than 30" Thick

LIGNITE
More Than 30" Thick
Less Than 30" Thick

Finding these deposits was joy without measure;
to him it was hunting for deep buried treasure.
Without the right data, it'd be difficult to know
exactly where a well or mine needed to go.

He wanted to do more to help his hometown,
so he founded a company in that small town.
Its purpose to help other tribes understand
how to develop resources on their own land.

The power to find resources and help others, too—
it's amazing what **mapping data** will do.

VERA RUBIN

GREGOR MENDEL

Scientists use diverse methods to explore;
doing their work on land, sea, and shore.
Despite all their efforts, at the end of the day,
without the right data, there's not much to say.

John Snow used data to stop a deadly disease,
and Gregor Mendel got lots of data from peas.
Data helped Vera Rubin make a strong case
for the existence of dark matter throughout outer space.

Data supports conclusions; it can change people's minds;
it is used to build theories that help humankind.
Scientists all along have known this to be true:
Data is powerful! Now, what will yours do?

JOHN SNOW

You can collect your own data! Here's how:

1 First, you need to decide on a question. You can find inspiration for questions in many places: at school, at home, and outside. Some questions you might like to investigate include the following: What type of liquid will freeze the fastest? What types of insects can be observed on our playground? How does a plant change over its life cycle?

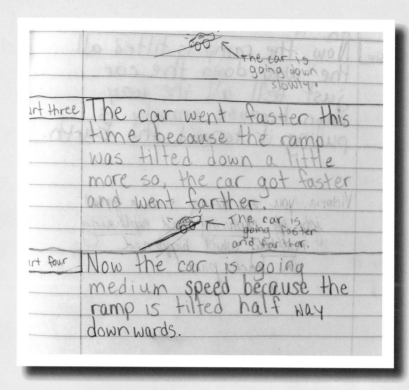

2 Having a question will help you decide what kind of data to collect. For some questions, you can sketch and write down observations. To answer other questions, you might need to conduct an experiment to collect data. You can use tally marks to record some kinds of data, but for others, you will need to measure. You can use different tools to measure time, distance, or weight. A data table, a special kind of chart, can help you keep your data organized.

3 Once you've collected your data, you need to analyze it. If your data is sketches and observations, you might compare them to find similarities and differences or how something has changed over time. If you have collected numbers and measurements, you can use math to analyze them. You could find out how much longer it took for one liquid to freeze by subtracting, or you could find the mean height of your plants. Making graphs is also a great way to show your data and see patterns.

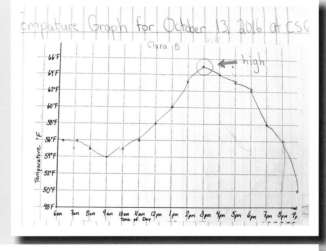

4 Finally, you can draw a conclusion about the data you collected. A conclusion tells what you learned from the data. It answers the question you asked in step 1. Sometimes, we call this conclusion a *claim*. No matter what you call it, though, your data needs to support it. We often call the data *evidence*, and it might make you think of new questions—which you can answer by collecting even more data!

Note on usage: We've chosen to use *data* as singular rather than plural throughout this book to make the text more accessible to young readers.

The scientists profiled in this book are a diverse group of men and women who have studied many different branches of science throughout history. Learn a bit more about them here.

Alhazen (Ibn al-Haytham) (965–1039) was born in Basra, Iraq, but spent most of his adult life in Egypt. His work spanned several fields: astronomy, mathematics, medicine, and optics. Alhazen was the first scientist to develop a hypothesis based on his observations of light and then devise an experiment by which to test this hypothesis. The simple experiment described in this book provided irrefutable proof of how vision works, disputing the widely held belief (at that time) that light rays emanated from the human eye and illuminated objects.

Maria Sibylla Merian (1647–1717) was a naturalist, artist, and entomologist. Encouraged to paint from a young age, she became interested in the caterpillars and butterflies she collected for her still-life paintings. She began keeping insects so she could study the changes they underwent throughout their life cycles. Later in life, she traveled to South America on a scientific expedition, something quite unusual for a woman at the time. Merian's work is characterized by incredible detail in both illustration and description. She depicted each stage of an insect's life cycle in conjunction with its food source, providing a great deal of evidence in support of the concept of metamorphosis and refuting the commonly accepted idea of spontaneous generation.

Dimitri Mendeleev (1834–1907) was a Russian chemist best known for creating the Periodic Table of Elements. Inspired by an international chemistry conference, he chose to arrange elements in order by atomic weight. The table was a success, despite other scientists' similar work, for two reasons: In arranging the data in the table, Mendeleev was able to (1) show that the atomic weight of several elements had been measured incorrectly and (2) predict eight new elements and their properties before the elements were even discovered. Over a hundred years and many new elements later, the periodic table still is the basis for understanding elements and their properties.

Carlos Juan Finlay (1833–1915) was a Cuban physician who devoted much of his spare time to studying malaria and yellow fever. Finlay noticed that outbreaks of these diseases roughly coincided with the rainy season, and he determined that patients who suffered from these diseases all reported having mosquito bites. This led him to conclude that mosquitos were the means of transmitting the diseases. Although he presented many papers on the topic, his ideas were not accepted for several decades, until U.S. Army surgeon Walter Reed led a commission on yellow fever. The commission accepted Finlay's research, which led to the eradication of yellow fever in Cuba and Panama, largely through methods proposed by Finlay himself.

Nettie Stevens (1861–1912) was an American geneticist who studied chromosomal inheritance and sex determination. Stevens found that mealworm beetles produced two kinds of sperm, one with a large chromosome, and one with a small chromosome. When the sperm with the large chromosomes fertilized eggs, they produced female offspring, and when the sperm with small chromosomes fertilized eggs, they produced male offspring. In contrast, female beetles produced eggs with only large chromosomes. This data led Stevens to conclude that males were responsible for the determination of the sex of offspring.

Ruby Hirose (1904–1960) was a biochemist and bacteriologist of Japanese American descent. She was the first second-generation Japanese American to graduate from her high school, and she eventually received a PhD in chemistry from the University of Cincinnati. Working at William S. Merrell Laboratories, she researched hay fever and other vaccines. In 1940, Hirose was among ten women recognized by the American Chemical Society for accomplishments in chemistry. She made major contributions to the development of vaccines against infantile paralysis, also known as polio.

Marie Tharp (1920–2006) earned degrees in geology and mathematics during World War II, when men were mostly absent from university settings. She found a job in Columbia University's Lamont Geological Laboratory, where she worked with a group of people known for pushing boundaries of scientific research. As a woman, Tharp was not allowed on the research vessels that collected sounding data to determine the depth of the ocean; instead, she plotted measurements by hand and collaborated with geologist Bruce Heezen on a map of the ocean floor. While Tharp's painstaking work led to the discovery of the mid-Atlantic ridge, a finding that provided support for Alfred Wegner's theory of continental drift, Heezen published and was given credit for the work.

Marie Daly (1921–2003) was the first African American woman to receive a PhD in chemistry. She specialized in biochemistry, studying how proteins are synthesized in the body and what causes heart attacks. Daly, assisted by Dr. Quentin Deming, was the first to connect high cholesterol and clogged arteries, deepening the understanding of the role of food in the health of the heart and the entire circulatory system. Daly also took an active role in encouraging African American students to study science, establishing a scholarship fund for students at Queens College in her father's name.

Russell Stands-Over-Bull* is a member of the Crow Native American tribe and grew up in a tribal community on the Crow Indian Reservation near Billings, Montana. He studied geology at Montana State University and the Colorado School of Mines, eventually earning his PhD while working for the oil industry. In 2001, he returned to his hometown and founded Arrow Creek Resources, a company dedicated to helping Native American tribes develop their mineral and energy resources. He is also an adjunct professor at Montana State University.

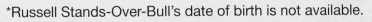
*Russell Stands-Over-Bull's date of birth is not available.

Image Credits

p. 8: Public domain. *http://againstthemodernworld. blogspot.com/2010/11/eye.html.*

p. 11: Public domain: *https://commons.wikimedia. org/wiki/File:Maccai_Maria_Sibylla_Merian_1705_ plate_VI.png.*

p. 13: Public domain. *https://commons.wikimedia. org/wiki/File:Periodic_table_by_Mendeleev,_1871.svg*

p. 19: Smithsonian Institution Archives. Image # SIA2008-3224.

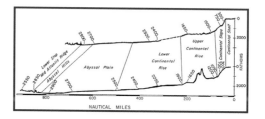

p. 21: Public domain. From "The Floors of the Ocean: 1. The North Atlantic" by Bruce D. Heezen, Marie Tharp, and Maurice Ewing. *https://www.gutenberg. org/files/49069/49069-h/49069-h.htm.*

pp. 28-29: Photographs courtesy of Bondi Photography, LLC.

About the Author

Jessica Fries-Gaither is an experienced science educator and an award-winning author of books for students and teachers. Her first children's book, *Notable Notebooks: Scientists and Their Writings,* was named a 2017 Outstanding Science Trade Book for Students K–12 and was read aloud aboard the International Space Station through the Story Time From Space program. Jessica teaches elementary science at the Columbus School for Girls. She lives in Columbus, Ohio, with her husband and three dogs.

Photograph courtesy of Bondi Photography, LLC.